Black Sunshine

The Story of Coal

Black Sunshine

The Story of Coal

Olive W. Burt

illustrated with photographs and drawings

Julian Messner New York

Design by Madeline Bastis and Friends, Inc.

Library of Congress Cataloging in Publication Data

Burt, Olive Woolley
    Black Sunshine: The story of coal.

    SUMMARY: Describes the formations of coal, how it is
mined, and the lives of the miners.
    1. Coal—Juvenile literature. [1. Coal. 2. Coal
mines and mining] I. Title.
TN801.B87   553.2   77-10874
ISBN 0-671-32847-6

Dedication

To John Gabriel Imlay
and
Deven Aleece MacNair.

Acknowledgements

I wish to express grateful appreciation to the many state historical societies, picture services and government agencies that so cheerfully supplied information and illustrations for this book. Especially helpful were Margaret Lester, Utah State Historical Society; James R. Arnold, Pennsylvania Historical and Museum Commission; Candace Greene, Stovall Museum, University of Oklahoma; Jack Bollow, U.S. Steel Corporation; William W. Corbitt, American Electric Power Service Corporation; Linda Kaufman, Montana Power Company; Dan B. Shields, West Virginia Coal Association, and the National Coal Association. And special thanks to Claude P. Heiner, international coal mining consultant, who from the first took a generous interest in the project and shared his vast knowledge of the coal industry with me.

Contents

Coal heated yesterday's homes: a furnace in the basement, and flues, or pipes, to carry the heat to every room in the house. Few homes today use coal.

Introduction— An Unseen Worker

Well, I just don't believe it," said Mark, pushing his textbook aside.

His mother looked up from her magazine. "What now?"

"It says here that every American uses more than twelve pounds of coal every day. You don't. I don't. I've never even seen a lump of coal. How can I be using twelve pounds of it—*every* day?"

"Because you use things that come from coal, or are made possible by coal," answered his mother. "The electric light, our television, the refrigerator—most things that use electricity depend on coal. Coal is needed to make steel, and we use things made of steel, like our car, and stove, and——Oh, you can think of a lot more, can't you? Even our clothes are manufactured with steel machines, and by electricity."

Mark shook his head. "I'd just like to see some coal! I'd like to know what it is——"

Less than 50 years ago, most children knew

what coal looked like. They knew it was black and dirty, and they knew what it did. It heated their homes and cooked their food. They might grumble when they had to fill the coal bucket and tend the furnace, but they enjoyed the warmth.

Then, gas and oil replaced coal in the home. But coal is still being used, working unseen in many ways, just as Mark and his mother were discussing. And now, as gas and oil supplies are running low, the world is turning more and more to coal for energy.

What is coal, where does it come from? How is it used? How much coal is there in the world?

The story of coal starts millions of years ago. . . .

What Is Coal?

Millions of years ago, swamps and forests covered much of the land. The climate was hot and moist. Plants—ferns, grasses and mosses—grew to enormous size.

The sun poured its energy into the leaves and stems of the plants. The plants then changed this energy into starches and sugars, which they used for growing. When the plants died, still loaded with starch and sugar, they fell into the swamp, and lay there, rotting. Other plants grew over them, and, in their turn, also died and fell.

As thousands of years passed, the earth's climates changed, and the land shifted. Mountains were pushed up, valleys were formed. Rains and

An artist's idea of how most of the earth must have looked during a Coal Age—giant ferns and mosses in swampy land.

melting snow flowed down the mountain sides, carrying *sediment*, or sand and mud into the valleys and low-lying swamps. The weight of the sediment and water pressed the decaying swamp plants into a soft, moist, spongy mass. As the centuries passed, some of the spongy masses were pressed so hard they became solid, like rock. They became coal. The changing of plant material into a mineral is called *fossilization* and coal is called a *fossil fuel*.

This happened in many places in the world. It happened at different times in the past, and is happening even now. But the most coal was formed in one period of time, millions of years ago, called the *Carboniferous Period*, or Coal Age.

A fern—Alethopteris—of the Coal Age, preserved intact in rock, and found in a coal mine.

The Coal Age, and each coal-forming period, was followed by times of climate changes. In some of these times, it became too cold or too dry for swamp forests to live. They disappeared. But rivers and rain and melting snows kept bringing in more sediment. The old swamp forests were buried deeper and deeper underneath these heavy layers of sediment. The great heat and pressure inside the sediment turned it into rock, covering the layer of coal.

Sometimes the climate would get hot and moist again. The swamp would then return, and start a new *cycle* of plant growth and decaying. This would, over many centuries, become another layer of coal. And again, climates would change and sediment would turn into rock.

Each layer of pressed plant material that became coal is called a *seam* and each seam is separated by layers of rock.

The amount of heat and pressure above seams of coal was not always the same everywhere. Wherever there was less heat and pressure, the spongy matter did not harden as much. In its softest form, this matter is called *peat*. Peat is the first stage of coal. It has not had all the moisture squeezed out of it yet, so it does not burn as well. But it can be used for fuel when dug up and dried out.

Peat under more heat and pressure over a longer time becomes *lignite*. Lignite is the second stage of coal. Lignite burns better than peat, but it still has some moisture left in it.

Lignite under more pressure and heat, and over more time, is turned into *bituminous coal*. Bituminous coal, or soft coal, is the third kind of coal.

Peat

Lignite

Bituminous Coal

Anthracite

There is more of bituminous coal than any other kind.

But bituminous coal is not the last stage of coal formation. If the coal stays in the rocks even longer, and with still more heat and pressure added, it becomes *anthracite*, the hardest and best kind of coal. Anthracite burns hotter and has less smoke. But since so much time, heat, and pressure are needed to form anthracite coal, there is not very much of it.

And in a few places in the world, anthracite was put under still heavier pressure and higher heat, for a longer period of time, and it turned to—diamonds! Diamonds are precious. But today, coal is almost as valuable. The world badly needs the energy released by the sunshine of thousands of years ago, the black sunshine of coal.

Coal
And The Growth
Of The Country

The Chinese burned coal over 3,000 years ago. Europeans discovered the use of coal a little later. The Europeans who came to America knew about coal, but they preferred to burn wood from the great forests around them.

The newcomers found that some Indians had long known about coal. Some eastern tribes polished it and used it for jewelry. In the West, the Utes had discovered by accident that the rock

would burn, but they made no use of this knowledge. Some Pueblo Indians, however, had learned to use coal in their blacksmith forges, and to heat their pottery ovens.

In 1679, Father Louis Hennepin, a French priest and explorer with La Salle, published information of his discovery of coal near the Illinois River. At Racine, West Virginia, a marker honors John Peter Salley for discovering, in 1742, the rich coal fields of that state.

But there was little need for coal in those days. Wood was burned for warmth, unless someone had some coal that was easy to get at on his property. Then he might use it, and perhaps sell some to neighbors.

LEFT: An old drawing of Zuni Indians of New Mexico using coal in their blacksmith shops and to heat the ovens in which they glazed their pottery. BELOW: To honor the discovery of coal and its importance to the state, this house made of coal stands on a main street in Middlesborough, Kentucky.

There were few industries in pre-Revolutionary times. Great Britain supplied its colonies with manufactured goods. The colonists were not permitted to make goods in competition with British manufacturers.

But after the American Revolution, the people of the new United States had to establish their own industries and build a nation. They moved westward across the mountains, cleared the great forests, built roads, set up farms, and started factories to make goods.

All this took large amounts of fuel, and people began to turn to coal. Soon, mines were dug to get to deeper seams, and coal was sold far and wide.

In the West, the early settlers found large areas that had few trees. They knew they could not afford

to burn wood, as the eastern colonists had done. Brigham Young told his followers settling in Utah, "We must find coal." A thousand-dollar reward was offered to the first person who found a sizeable coal field. So the Utah pioneers really hunted for coal, and found it in several parts of the territory. Later, deposits were found also in Colorado, Wyoming, Montana, New Mexico, and the Dakotas.

Coal fed the furnaces to make the iron and steel that factories needed. Railroads began to use steel rails, which were safer than wooden ones. Locomotives began to burn coal instead of wood. Coal was used in making guns for the Civil War of 1861-1865. And coal supplied the fuel for the great industrial expansion afterward.

Factories sprang up everywhere. But they needed workers. The poor people of Europe

Coal areas in the United States.

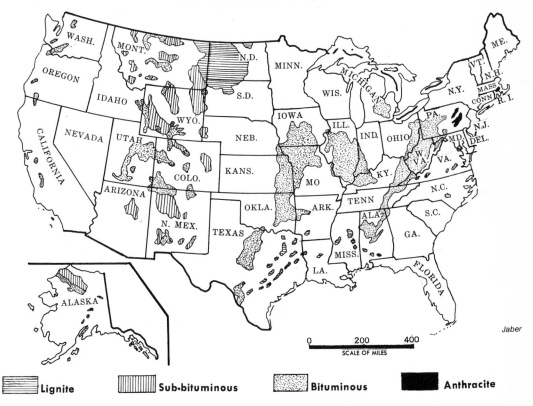

Jaber

Lignite	Sub-bituminous	Bituminous	Anthracite

learned that there was plenty of work in America. Immigrant ships crossed the oceans in increasing numbers. The newcomers went into the factories. They helped lay the steel tracks for the railroads. They went into the coal mines.

Boys as young as 9 or 10, small enough to get into narrow places, also went into the mines. They helped dig the coal that kept the trains moving and the factories roaring.

Coal became the most important source of heat and electricity. Coal heated the water that made the steam to turn the giant wheels which generated electricity. Locomotives used coal to make the steam that turned their wheels. Factories used coal for steam and heat. Homes used coal in stoves and furnaces, for cooking and heating.

Then, suddenly, coal had a rival. On an August afternoon in 1859, at Titusville, Pennsylvania, a driller named Edwin Drake struck *oil*, a liquid fossil fuel. A few years later, underground supplies of *natural gas* were also discovered. Natural gas is also a fossil fuel formed like oil, but it is not a liquid. It is a gas, like air, and has to be kept in tanks under pressure.

People and industries began to try the new fuels. But coal continued to be used in large amounts.

Then another way was found to make electricity. Falling water turned giant wheels that generated electricity in *hydroelectric power plants*. Although generating electricity remained one of coal's most important jobs, its use went down.

And another way was found to run engines. A young man named George Selden invented an engine that ran with gasoline. Gasoline is made from oil. The new gasoline engines started taking over some of the jobs that had been done by coal-burning steam engines.

Coal was being used less and less.

By the 1930's, oil-fired or natural gas furnaces were replacing coal furnaces for heating homes and factories. They made cleaner and steadier heat—there was no coal to shovel, no ashes to take out, no fires to be watched.

The production of coal went down even further. Only one of coal's most important jobs has not

White-hot coke pours from a coke oven into a quenching car, where it is cooled. Making coke is one of coal's biggest jobs.

been replaced by another fuel. Nothing but coal can be made into *coke*. Coke is coal that has been baked in an air-tight oven. Coke makes the hottest fires, necessary for making steel out of iron ore.

But from World War II until today, the coal mining industry has shrunk. Oil and natural gas can do almost everything that coal can do. They burn cleaner than coal. And they can be transported in pipelines for less money than it costs to transport coal by trains. Mines closed down; miners were without work.

But today, the picture is changing again. The oil and natural gas supplies are giving out—and once more the country is looking to coal for energy.

Into The Earth
For Coal

Ten-year-old Danny Tamrowski knows about coal. His father, Peter Tamrowski is a coal miner. The family lives in a Pennsylvania coal town that saw hard times for a while. But things are much better now.

Five days a week, Danny's father drives from his home, through the town, and up a steep road to the top of a hill. He parks near a large wooden building and goes inside. Other cars are parked there, and more are arriving.

The men greet each other and go inside to the *dry room*. This is a large room, with shower stalls.

Danny Tamrowski likes to put on his grandfather's old hard hat and pretend.

ABOVE: A modern dry room is spacious and clean. The newly-washed clothes are hung near the ceiling, to leave uncluttered space for the men to change their clothes. In the rear are bright, clean showers.

LEFT: Miners in a large modern mine wear what is almost a uniform: coveralls, hard hat, hard-toed shoes. Everything is snug and tight with no dangling strings or flapping material to cause accidents.

25

An early mantrip car.

Here they remove their street clothes and put on heavy coveralls over a woolen shirt. The shirt must be warm, because it is always cool in an underground mine. They wear hard hats for safety. Today's hats have a battery-operated lamp in front. Years ago, candles were used.

A group that works together is a *section crew*. Peter Tamrowski's section crew walks down a hall to an elevator, which they call a *cage*. It drops the crew straight down through the mountain.

It is dark in the elevator shaft and the men cannot see what they are passing. But they know that wherever the shaft cuts through a seam of coal, tunnels lead off along the seam. They are going to work in a tunnel at the 1200-foot level. That is almost as deep as the tallest buildings in the world are high.

When the cage stops, the miners climb into iron *mantrip* cars, a little train that takes them along a tunnel dug out of the coal. They get out when they reach the section where they will work today, a room carved out of solid coal.

But this room is not dark, with black walls and ceiling. No, it is white. The whole inside of this area has been sprayed with powdered limestone. This keeps down the dangerous coal dust and makes it easier for the miners to see what they are doing.

Steel ceiling props are set in place by electric machines. Floors, walls, and ceilings have been sprayed with powdered limestone.

One of the crew moves about, holding up a small metal box. This is a *methane detector.* Methane is a type of natural gas, nearly the same as the gas used to cook with. It formed along with the coal millions of years ago. In the open areas of coal mines, it is very dangerous because it explodes easily.

The detector will show whether there is methane gas in the area, or any other dangerous

Using a methane detector.

gas. If any is present, the crew will move on to another area until this one is cleared and safe.

In this large modern mine, most of the work is done by machines. Danny's father is a well-paid specialist, who operates an electric machine called a *continuous miner*. Danny's father can mine and load up to 12 tons of coal a minute with this machine. It has sharp rotary blades that slice into ceiling or wall to gouge out tons of coal. The shattered coal falls to the floor. The machine has steel arms that reach out and scoop up the coal, and drop it onto a conveyor belt. The belt carries it to the cage where it is hoisted up the shaft to the surface.

Peter Tamrowski operates a continuous miner.

29

The steel arms of the loader scoop up the broken coal and dump it into a shuttle car or onto a conveyor belt which will take it to the cage. There it will be hoisted to the surface.

At noon, the men do not leave the mine. They gather in the *grub pit*, and talk and laugh as they eat from their lunch boxes.

The men do not leave the shaft until a whistle blows to tell them their day's shift is over. Then they ride the mantrip cars to the cage and are whisked up the shaft. In the dry room again, they take off their dirty coveralls and toss them into automatic washing machines. By the time the men have showered and dressed, their coveralls will be ready to be hung up to dry.

Peter Tamrowski remembers when there were no showers and electric washing machines. Wives washed the miners' dirty clothes by hand. And the miners bathed as best they could in big metal tubs of water. Every miner's shack had a tub hanging on the back porch.

Danny's grandfather, Claus, is a retired miner. He was a baby in his father's arms when the family came from Poland 75 years ago. Many coal miners from Poland had settled in this area where they had found work in the mines.

In earlier times and even today in smaller mines, the men clean up at home. Here they are seen walking out of a mine at the end of a shift. It's dirty work!

Claus Tamrowski tells his grandson, "I had a pick and shovel and my own muscles. That was all. We dug our tunnel into the mountain. We walked in and brought out the coal on our backs—12 to 16 tons a day." Claus's son Peter mines that much coal with his continuous miner in one minute.

Claus was just Danny's age, 10, when he went into the mine as a *breaker boy*. These youngsters worked in close places where men could not fit. And they picked out slate and rock as the coal moved past them along a wooden chute. "Even that is done by machine now!" Claus says, "and that is good!"

An artist's sketch of young boys working in a mine in the old days.

Diagram of a shaft mine.

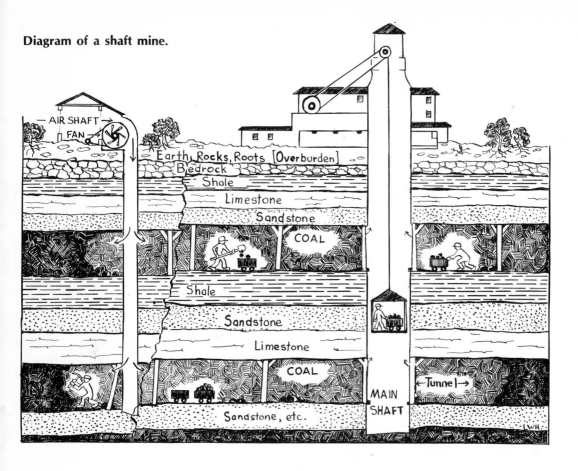

AIR SHAFT
FAN
Earth, Rocks, Roots [Overburden]
Bedrock
Shale
Limestone
Sandstone
COAL
Shale
Sandstone
Limestone
COAL
Sandstone, etc.
MAIN SHAFT
Tunnel
(WH)

Peter Tamrowski is working in a *shaft mine*. The coal lies very deep and is reached by a shaft. When the coal can be reached by digging a tunnel through the side of the hill, straight to the seam, the mine is called a *drift mine*. The crews do not need a cage to reach the tunnels, but go straight in with mantrip cars. If the coal lies below the base of a hill, or under a lake or sea, the tunnel must be slanted, or sloped, down from the surface. This is a *slope mine*.

Diagram of a drift mine.

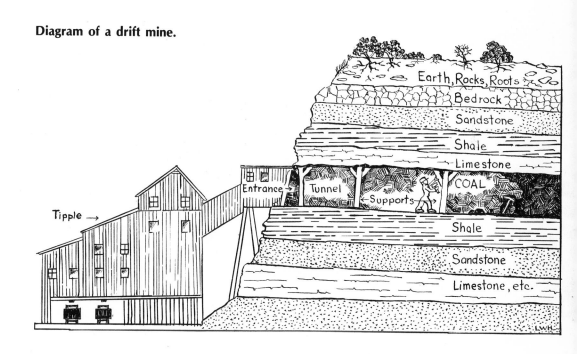

Earth, Rocks, Roots
Bedrock
Sandstone
Shale
Limestone
Entrance → Tunnel   COAL
← Supports
Shale
Tipple →
Sandstone
Limestone, etc.

LWH

Claus and Peter Tamrowski have always worked in underground mines. Several years ago, Peter's brother, Matt, moved to Ohio to work for a company whose coal beds lay near the surface. The coal could be reached by stripping away the shallow layer of earth and rock, called *overburden*, that covers it. This is *surface* or *strip mining*.

Today, more than half the coal mined in the United States is obtained by surface or strip mining.

Strip mining.

It is quicker, easier and cheaper than is underground mining. Once it was safer, too, but that is no longer true.

And surface mining can scar the land. The *bulldozers* used to clear away the overburden, and the *draglines* (very large cranes) that scoop out the coal, leave open pits and piles of rock debris.

The Navaho Indians of the Southwest were shocked at the idea of ripping up their Mother Earth to get coal. They made such effective protest that a plan for mining desert coal was abandoned.

When Peter talks about this, Matt reminds him that the land need not be ruined. He tells of the recreation parks being established on strip-mined land in Ohio, Kentucky and West Virginia, and the Montana land seeded with grain, grass and trees. Such *reclaimed land* is being put to good use in many parts of the country.

There are laws that strip-mined lands must be reclaimed, but they are not always enforced. However, the federal and state governments are getting more strict. And stronger laws are being passed. Some companies still object because they want to keep the cost of coal down. And they do not want to be slowed down when coal is urgently needed. But most people agree that the most important thing is that the land not be scarred forever.

Camping and biking on reclaimed strip-mined land in Ohio, now a park.

The Coal Miner— Then And Now

Although Claus Tamrowski likes to poke a little fun at his son Peter for using machines to do the hard work of mining, he never jokes about the changes that have made the life of a miner safer and healthier. Claus remembers how it was when he first came to America. He remembers the work and the suffering it took to get these changes. He helped bring these better days about.

When Claus was a young man, the mine owner was like a king. He owned not only his mine, but everything connected with it—the land, the town, the workers' houses, the store, the church, almost the miner himself.

A typical early coal town, built in Whitmore Canyon, Utah, about 1900. It developed into the rich coal-producing town of Sunnyside, where at one time more than 1200 miners were digging coal.

It was common for a coal miner to work from 12 to 15 hours a day, six days a week, for as little as two dollars a day. Of course, in those days (about 1900), a loaf of bread cost only three cents. Still, the pay was very low.

Then the miner was not even paid in money, but in metal tokens and slips of paper called *scrip*. Before the miner received this scrip, the owner had taken out the rent for the shack where the miner's family lived, a "medical fee" for a doctor the miner seldom saw, and any other amount the mine owner could think of. And the scrip had to be spent at the company store, where miners were charged more than they would be at other stores.

Boys went into the mines as soon as they could, usually at ten. They grew up without an education. Mining was all they could do, whether they liked it or not.

The miner was always in debt because what he earned was never enough to meet the mine owner's demands and also feed and clothe his family.

Little attention was paid to the miner's safety or health. When a miner contracted a disease or was injured in an accident, he was laid off. There was no more work, no more money for him.

No one seemed to care. When things became unbearable, around the beginning of the twentieth century, the miners themselves decided to do something.

They had only one weapon: to go on *strike*. They would stop digging coal. To go on strike meant that there would be trouble. The mine owners had money and power. They could call for help from police and from United States troops. It meant violence.

In order to make a strike work, all the miners had to band together. Organizers went about the coal camps urging the miners to form *unions*, or groups. Complaints and demands from a union meant more than the complaints from a few workers. As an old song puts it:

The boss won't listen when one man squawks,
But he has to listen when the Union talks.

RIGHT: Pinkerton Detective Robert J. Linden in charge of Coal and Iron Police putting down rioting strikers of the Mollie Maguires, a Pennsylvania coal miners' union, in 1875.

Separate unions were organized here and there. They got some action, but were not always successful. The miners saw that they must form a stronger organization. Many small unions joined together and found that this helped. Finally, in 1890, a national organization was formed, The United Mine Workers of America (UMWA).

By 1920, with a powerful man, John L. Lewis, as president, the UMWA was able to bring about many reforms in the coal mining industry.

But there was a lot of violence, as mine owners at first fought the unions. There were many strikes. Sometimes the owners threw miners and their families out of the company-owned houses, forcing them to live in tents. *Strike-breakers* were brought in—men who were willing to work for the old pay, and without reforms. Sometimes miners had to fight armed guards, hired by the owners to protect their strike-breakers.

The unions won most of their reforms. Safety standards were established. Scrip was abolished. Pay raises were given. A pension system was set up to allow miners to retire with an income after many years of service. Another system was set up to pay workers who had been injured or disabled in the mines.

The state and federal governments helped miners to get these reforms. They passed laws to enforce standards and to force both the companies and the unions to obey the rules they agreed to in their talks.

Even with the union's gains, the miner's life was hard. He was often out of work. Some of the small coal companies were forced out of business by the increased labor costs. Others closed down as the demand for coal lessened. Some places turned into ghost towns as families looked for work elsewhere. Most mine towns remained very poor.

Today, the miner's life has improved greatly, and so have work conditions. Planned communities of homes, schools, recreation facilities, and libraries have taken the place of the old shacks of the company-owned coal camp. Many of the homes were built by the mine owners and sold to the miners at reasonable prices.

At the present time, the American coal miner produces more coal a day than does any other miner in the world. American miners have a higher standard of living than miners in other countries. They often own their own homes. A miner's children attend good schools. Women may stay at home to care for the family, or they may work. A woman may even be a coal miner herself. And what a change that is!

From its very founding almost 100 years
ago, Colstrip, Montana was known as a
model coal camp. Today it keeps that
honor, as can be seen from these
attractive homes and "tot lots"—play
areas for small children.

It used to be believed that if a woman entered a coal mine, something awful would happen. Stories were told of cave-ins, explosions, and floods that came right after some woman entered a mine. But in 1973, the Bethlehem Steel Corporation hired two mothers as coal miners. The women dug coal—and nothing happened. Three years later, that company was hiring 88 women, 80 of whom dug coal. Today, there are some 250 women who don copper-toed boots and hard hats and go down into the earth to bring out coal.

Mrs. Diana Baldwin and Mrs. Anita Cherry—the first women coal miners employed by Bethlehem Steel Corporation.

Dangerous Work

Going into the earth for coal has always been dangerous, particularly in the early days. However, every year the number of accidents and sicknesses in the coal industry grows smaller. The federal government and the governments of coal-producing states require that strict safety measures be established in every mine. The diseases that once affected almost every coal miner have been studied, and better ways of preventing and treating them have been found.

Coal dust is an ever-present danger. Breathing it in over long periods of time can cause *black lung*, a disease that causes many deaths.

Coal dust is also an explosive. It can be set off by almost any spark or flame. The coal dust level is kept down by spraying with powdered limestone or with carefully controlled dampening, so explosions are not as common as they used to be, but they do still occur. Nothing that will make a spark or flame is permitted in an underground coal mine.

Cave-ins from explosions or falling over-burden once were very frequent. Today, steel roof props are used and bolted to the ceilings. This has done much to make cave-ins less common.

An early mine with wooden props—dangerous because they were too weak and a hazard in case of fire. Flames would leap from post to post, spreading rapidly through the mine.

The use of steel props and the bolting of the ceiling to the rock above have done much to lessen the number of cave-ins and fires.

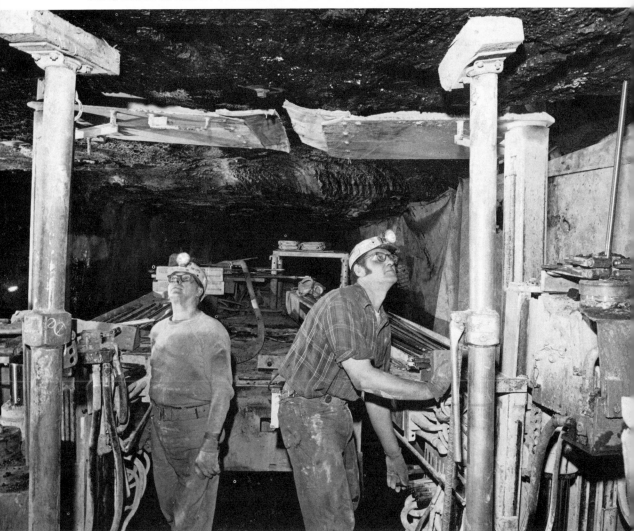

Another danger is underground water which may suddenly flood the tunnels, drowning or trapping miners.

Two gases can sometimes build up in underground coal mines. One is methane, the other carbon dioxide. They are both colorless, odorless and tasteless—but deadly. In the past, miners would take small birds, usually canaries, down into the mines with them. If there was the slightest amount of either gas in the mine, the birds would fall over dead. The men then had a chance to escape. Today, there are scientific detectors, like the one used by Peter Tamarowski's crew.

Motorized machines have made underground mining easier and safer in some ways than it used to be. But these machines have created dangers of their own. Operating powerful machines in cramped spaces requires great skill and watchfulness. The slightest misstep could cause an accident.

Fresh air is an excellent safety device. A well-managed mine has fresh air forced through tunnels and work areas. If the air is kept moving, the men can work and breathe in comfort.

Iron shuttle cars travel through the tunnels, collecting the coal. Notice the limestone-sprayed walls and ceiling.

Fresh air is the best
safety measure in an
underground mine.
Here two great fans,
mounted outside a
shaft, suck in fresh air
and force it down into
the tunnels.

Inside the mine,
flame-proof curtains
can be opened or shut
to control the flow of
air.

It used to be that surface mining was much safer than underground mining, but that is no longer true. There are no roofs to cave-in, no dust to explode or clog the lungs, and no flood danger. And there are no poison gases. But bulldozers that scoop up the overburden above a coal seam move tons of earth and rock. The great shovels swing out above work roads and workers. The draglines that

After the shovel has uncovered the coal, a dragline comes over and digs into the seam. The shovel at the end of the long arm swings the coal to a stockpile or a railway car or barge which may be waiting.

remove the coal are sometimes as tall as a 20-story building. They, too, swing tons of heavy material across wide spaces. Beneath them any living thing is in danger. In 1975, for the first time in our coal mining history, more men were killed working surface mines than died in underground accidents.

Every mine owner and every coal miner is aware of the dangers of mining. And most accept the responsibility for any threat to safety.

Instructions in use of safety equipment. The instructor is from the Bureau of Mines.

Most large companies hold classes in safety and proper use of equipment for newcomers to mines. Most have safety engineers on their staffs to study ways to avoid accidents and to educate employees in safety measures.

The United States Bureau of Mines was established in 1910, to study ways of preventing accidents and then to see that its safety measures were carried out. States also have mine bureaus, with inspectors to check on how well the mines in the states are following rules. If there are violations, a mine may be closed down until standards are met.

In most of the coal mining areas there are some mines that are illegal. They are small and hidden in forested slopes and operate secretly. These are called "wildcat" mines. They often operate unsafe equipment and do mining under dangerous conditions. Some of these mines do not have limestone rock dusting to keep down dangerous dust. They do not prop up their roofs properly. And they often employ people who are not trained in mine safety. People who need work badly sometimes accept these conditions.

Coal For The World

Mine-run, or *raw coal* is sold just as it comes from the mine—dirty, with bits of slate, rock, and other unwanted matter clinging to it. But usually coal is cleaned and sorted before it goes to market.

Railroad cars, trucks, or conveyor belts, carry the coal from the mine to the *preparation plant*, to be made ready for use. At an anthracite mine the preparation plant is called a *breaker*, because hard anthracite must be broken up into usable chunks. At a bituminous mine, the preparation plant is called a *tipple*. In earlier times, the coal cars were tipped over to dump their load. This gave the building its name.

Whatever it is called, the big, factory-like buildings house machines that wash and dry the coal. Then, in the more up-to-date plants, the cleaned

The Huber breaker of the Glen Alden Coal Company near Wilkes-Barre, Pennsylvania. Thousands of tons of anthracite are prepared for market in this breaker, which is one of the largest and most modern in the world.

coal is run over a set of screens. These sort the coal into 7 different sizes. The finest pieces fall through the small openings in the first screen. The next size falls through larger spaces in the next screen. This goes on until only the biggest chunks are left.

From time to time samples are picked from under the screens. These go to the laboratory. There they are tested to see how much unwanted matter they have. There are some impurities in all coal. The coal with the least amount of impurities burns best.

Today, the fastest and cheapest way to transport coal is by railroad. More than two-thirds of the coal mined in the United States is sent to market in special open railroad cars called *hoppers* or *gondolas*. A train of coal cars, called a *unit train*, can move as much as 15,000 tons of coal a trip. The coal mined each year in the United States could fill a unit train long enough to reach almost three times around the earth.

Trucks are used to carry small amounts of coal to nearby markets. And for countries across the ocean, the coal is taken to a seaport by train, barge, or truck. There it is loaded into a coal ship, a *collier*, to complete its journey.

A unit train of coal.

Transporting coal adds to its cost. To bring the price down, companies are trying to find cheaper ways of moving coal. Bituminous coal is sometimes shipped as *slurry*. Slurry is made of crushed coal mixed with water to make it flow. It is about as thick as hot cereal mixed with milk. This can be sent through pipelines. The buyer dries out the water and has a fine, clean, powdered coal. Pipeline delivery is fast and cheap. But sometimes it is difficult to get permission to build a pipeline. States and landowners often object to having pipelines cross their land. Nevertheless, there are several slurry pipelines in operation in the United States, Poland, China and a few other countries.

Some companies are beating the high cost of transportation by cutting down on the distance coal must travel. They are building their electric power plants or steel furnaces or factories close to a coal mine.

The use of coal is increasing rapidly as our reserves of oil and natural gas begin to run low. More factories are converting from oil to coal. The Federal Energy Administration is ordering some power generating plants to convert also.

Coal is being used up faster than nature can make it. It used to be said that the United States had enough coal to last 500 years. Now experts are saying, "Well, maybe 300 years!" What will they say a few years from now?

Coal Today
And Tomorrow

More than three-quarters of all coal mined in the United States today is used to produce just two things: electricity and steel.

But, when coal is burned, it is not used up completely. Tar, ammonia, gas, oil and sulfur escape from the burning coal. For years these "leftovers" were a nuisance. They gummed up furnaces, sent poisonous fumes over the land and filled the air with smoke. Today, these "left-overs" are processed to give us more than 2,000 useful articles. These are called the *by-products* of coal.

Some of the things in which coal has been an unseen worker. Plastics, nylon, synthetic rubber, and more than 2,000 other items are made from, or by, coal.

Among these by-products are the big family of plastics, nylon, synthetic (man-made) rubber, drugs, paints, linoleum, varnishes and explosives. And the end is not yet in sight. Every day some scientist comes up with a new idea for a new product made from the "left-overs" of burning coal.

What will coal do in the future? The federal government's Office of Coal Research and the laboratories of the major coal companies have experts studying the future of coal. They have already found a way to turn coal into gas—*gasification*—by burning it underground. The gas produced can be used in place of natural gas.

Up to now, coal has been burned as a solid. Scientists believe that coal can also be made into a liquid fuel. This is called *liquefaction*. Liquid coal could then be used in place of oil-based fuels.

Some scientists are working on the idea of doing coal mining by remote control. An expert will sit in a control tower and push buttons that operate machines in the mines. With no operator present, the machines would dig, load and move coal faster than human beings could. Wouldn't Claus shake his head at that!

But scientists are also worried about burning coal. Coal smoke contains carbon dioxide and sulfur dioxide, which trap the heat of the sun. Over many years, this trapped heat would cause the climates of the earth to become much warmer. Farm lands would turn to desert, and Arctic ice would melt, raising the sea level and flooding coastal cities. To prevent this and other pollution problems, researchers are trying to develop a smokeless coal.

There are several processes which will turn coal from a solid into a liquid. Dr. R.E. Wood, a researcher at the University of Utah, shows some of this liquid coal which he and his team have developed. Dr. Wood and others believe that someday it may take the place of gasoline and oil, to run engines that require a liquid fuel.

Coal is the one source of energy that is abundant today. Other sources of energy, such as the sun (*solar energy*), the interior of the earth (*geothermal energy*), and the energy released by splitting the nucleus of an atom (*nuclear energy*), are all at hand. But, as yet, we don't have the means to put these energies to work for us. Nor do we know how to make the best and most economical use of them. But ways will one day be found. Until then, coal will be the unseen worker—the black sunshine—that helps to power our world.

Glossary

ANTHRACITE — The hardest type of commercial coal.

BITUMINOUS COAL — The most useful and most plentiful type of coal. It is also known as soft coal.

BLACK LUNG — A disease that miners get from breathing too much coal dust.

BREAKER — The plant that cleans and sorts anthracite coal.

BULLDOZER — A heavy tractorlike machine used in surface, or strip mining to clear away overburden.

CAGE — An elevator car in the shaft of a coal mine.

CARBON DIOXIDE — A colorless, odorless gas that comes out of coal when it is burned.

CARBONIFEROUS PERIOD, or COAL AGE — A warm, moist time millions of years ago, when most of the coal was formed.

COAL BY-PRODUCT — Any item made from coal or from the wastes and gases left when coal burns.

COKE — A type of fuel made by baking coal in an air-tight oven. Coke gives the best kind of heat for making steel out of iron ore.

COLLIER — A ship that carries only coal.

CONTINUOUS MINER — A machine with sharp blades and steel arms. It gouges out the coal, scoops it up and drops it onto a conveyor belt.

CYCLE — Any action or event that repeats itself at regular times, like day and night.

DRAGLINE — A giant crane with a scoop bucket that can scoop up many tons of coal at one time.

DRIFT MINE — A coal mine that is entered by a tunnel dug straight into the side of a hill.

FOSSIL FUEL — Any fuel that is formed from the remains of plants and animals. Coal, oil, and natural gas are fossil fuels.

GASIFICATION — The process of turning coal into a gas by burning it underground right in the mine. Only a small amount of air is let in, and the gas from the burning coal is piped out.

GEOTHERMAL ENERGY — Energy in the form of natural heat that comes

61

from deep down in the earth, and reaches the earth's surface through cracks, or through holes drilled down into the rocks.

GRUB PIT — A space or room in a mine, dug out of coal, where miners can safely sit and eat or rest.

HOPPER or GONDOLA — A type of railroad coal car.

HYDROELECTRIC POWER PLANT — A plant where the force of falling water is used to turn big wheels that generate electricity.

LIGNITE — The second stage of coal.

LIQUEFACTION — The process of turning coal into a liquid fuel which will burn just like oil.

MANTRIP CAR — The low iron cars that carry miners to the places where they will work.

METHANE — A colorless, odorless, explosive gas that is sometimes found in coal mines.

METHANE DETECTOR — A special portable machine that warns miners if methane is present in the area where they are working.

MINE-RUN COAL — Coal that is shipped to buyers without any sorting or cleaning in a preparation plant.

NATURAL GAS — A fossil fuel made up of a mixture of gases, but mostly methane. It is found with oil in some oil wells.

NUCLEAR ENERGY — The powerful energy that is released by splitting the nucleus (core) of an atom.

OIL — Also called petroleum, a dark liquid fossil fuel found in some places in the rocks deep underground. It is reached by drilling oil wells.

OVERBURDEN — The rock, soil, and debris on top of a seam of coal near the surface. See Surface, or Strip Mining.

PEAT — The soft, decayed remains of an ancient forest. It is the first stage of coal.

PREPARATION PLANT — A building where coal is sorted, cleaned and loaded into railroad cars. See also Breaker; Tipple.

RECLAIMED LAND — Land that was strip-mined and then leveled off and landscaped for recreation areas or seeded for farms.

SCRIP — Metal or paper money made by the mine owners and used to pay the miners. Mining scrip was good only in the owners' stores.

SEAM — A layer of coal between layers of rock.

SECTION CREW — A group of miners who work in one spot in the mine.

SEDIMENT — Sand, gravel, shells, and debris carried by running water and dropped on the bottom of the stream, lake or ocean.

SHAFT MINE — A mine that is reached by digging a hole (shaft) straight down into the ground, big enough to build an elevator in.

SLOPE MINE — A mine that is reached by digging a tunnel at a slant. Such mines are usually located in the rocks beneath the floor of a lake or ocean, or they are deep under the base of a hill.

SLURRY — Coal that is ground up, mixed with water and shipped in a pipeline. It has the thickness of hot cereal.

SOLAR ENERGY — The powerful energy that comes from sunlight in the form of heat.

STEEL — A very hard metal, made from iron and small amounts of other substances which are added to the mixture under high heat.

STRIKE — A refusal to work, in order to force an employer to meet demands.

STRIKE BREAKER — A person who is hired to try to make strikers quit or go back to work, or to work in their place.

SURFACE OR STRIP MINING — The mining of a seam of coal near the surface by taking off the overburden of rock, soil and debris, and loading the coal into trucks or trains.

TIPPLE — A preparation plant for bituminous coal. It sorts, cleans, and loads the coal into railroad cars or barges.

UNION — An association of miners or other workers for the purpose of helping each other get better wages and working conditions.

UNITED MINE WORKERS OF AMERICA (UMWA) — The largest and strongest miners' union, formed nearly 100 years ago.

UNIT TRAIN — A railroad train made up of coal cars only, and all going to the same place.

Index